CONVERGENCE SECURITY

Volume 1, 2016

CONVERGENCE SECURITY

Editor-in-Chief: Kuinam J. Kim, Kyonggi University, South Korea

Editorial Board
Abdul Raouf Khan, King Faisal University, Saudi Arabia
Alessandro Bianchi, University of Bari, Italy
Bo-Wei, Princeton University, United States
Jiqiang Lu, Institute for Infocomm Research, Singapore
Kennedy Njenga, University of Johannesburg, South Africa
Maicon Stihler, Federal Center for Technological Education of Minas Gerais, Brazil
Michel Toulouse, Vietnamese-German University, Vietnam
Nasir Uddin, Oracle Corporation, United States
Nathan Clarke, Plymouth University, United Kingdom
Oscar Pereira, University of Aveiro, Portugal
Sangseo Park, The University of Melbourne, Australia
Stephen Flowerday, University of Fort Hare, South Africa
Susan Mengel, Texas Tech University, United States
Wei-Chuen Yau, Multimedia University, Malaysia
Zeyar Aung, Masdar Institute of Science and Technology, United Arab Emirates

This journal is published in collaboration with the iCatse organization: www.icatse.org

Aims and Scope
Convergence Security can be defined as the security created by merging the physical and information security or the combining security with other non-IT technologies. *Convergence Security* Journal is an open access journal dedicated to the publication and discussion of research articles, short communications, and review papers on all areas of convergence security. The primary objective is to publish a high-quality journal with free Internet-based access for researchers and other interested people throughout the world. Papers published by the journal represent important advances of significance to experts within security field. Our aim is to encourage scientists to publish their experimental and theoretical results in as much detail as possible. There is no restriction on the length of the papers.

The scope of *Convergence Security* is theory, applications, and implementations of convergence security. This includes, but is not limited to:

- Physical and Information Security
- Vehicular Networks Security
- Aviation Networks Security
- Maritime Networks Security
- Smart Home Security
- IoT Security
- Smart Grid Security
- Healthcare Security
- Infrastructure Security

Published, sold and distributed by:
River Publishers
Alsbjergvej 10
9260 Gistrup
Denmark

River Publishers
Lange Geer 44
2611 PW Delft
The Netherlands

Tel.: +45369953197
www.riverpublishers.com

ISSN 2445-9992 (Online Version)
ISBN 978-87-93519-53-4

CONVERGENCE SECURITY
Volume 1, 2016

A Feature Selection Approach Based on Simulated Annealing for Detecting Various Denial of Service Attacks

In-Seon Jeong[1], Hong-Ki Kim[2], Tae-Hee Kim[2], Dong Hwi Lee[2], Kuinam J. Kim[3] and Seung-Ho Kang[4]

[1]School of Electronics & Computer Engineering, Chonnam National University, 77 Yongbong-ro, Buk-gu, Gwangju, Republic of Korea 61186
[2]Department of Information Security, Dongshin University, 185 Geonjae-ro, Naju, Jeonnam 58245, Republic of Korea
[3]Department of Convergence Security, Kyonggi University, 94-6 Yiui-dong, Yeongtong-gu, Suwon-si, Gyeonggi-do 16227, Republic of Korea
[4]Department of Information Security, Dongshin University, 185 Geonjae-ro, Naju, Jeonnam 58245, Republic of Korea
E-mail:{jis0755; kinston}@gmail.com

Received 25 February 2016; Accepted 27 March 2016;
Publication 16 April 2016

Abstract

Feature combinations affect network intrusion detection/prevention systems based on machine learning methods such as multi-layer perceptron (MLP) in terms of accuracy and efficiency. However, selecting the optimal feature subset from the list of possible feature sets to detect network intrusions requires extensive computing resources. In this paper, we propose an optimal feature selection algorithm based on the simulated annealing algorithm to determine six denial of service attacks (neptune, teardrop, smurf, pod, back, land). In order to evaluate the performance of our proposed algorithm, three well-known machine learning methods (multi-layer perceptron., Bayes classifier, and support vector machine) are used against the NSL-KDD data set.

Convergence Security, Vol. 1, 1–18.
doi: 10.13052/jcs2445-9992.2016.001

Keywords: Network intrusion detection system, Machine learning, Feature selection, Simulated annealing algorithm, NSL-KDD trustworthy, cyber-physical, identity, locator.

1 Introduction

An incomparable amount of changes is taking place in the methods and volume of information distribution owing to the widespread application of computers and the exponential increase of wired and wireless networks. Especially, the application of information and computer technology to various industries definitely contributes to increased efficiency and productivity. However, the wide spread application of computers and the increase of networks also elevated the incidence of malicious activities such as information leakage and intrusions, resulting in economic losses and difficulty with spreading information technology. To tackle these problems, varying methods have been proposed especially in the area of network-based intrusion detection systems (IDS). Almost all of the IDS deployed in networks are signature-based IDS and use a set of simple rules. Despite some advantages such as high confidence in detection and low false positive rate, signature-based IDS have limits on detecting unknown attacks and the need for expert knowledge to create signatures. For this reason, machine learning based IDS has attracted many researchers as an alternative IDS approach [1–8].

One of the most important factors in developing IDS based on machine learning methods is finding feature sets to characterize and describe attacks in networks. Although various features have been extracted from network packets and system logs and many others have been proposed, a public feature data set for objective and fair comparison between proposed IDS is required by reseachers. The KDD'99 data set was provided by MIT Lindon lab to fulfill this requirement [9]. Many researchers have used the KDD'99 data set to evaluate the performance of IDSs they proposed. However, owing to the disadvantages of the KDD'99 data set such as excessive data size, data redundancy and bias to certain attacks, the data set has limitations on using it without modification. To address the problems of the KDD'99 data set and provide a data set that can be used to carry out objective and fair performance comparisons, the NSL_KDD data set was proposed by Tavallaee et al. [10, 11].

Both the KDD'99 and NSL_KDD data sets use a total of 41 features to characterize and describe various attacks. However, a total of 41 features is not suitable as a descriptor for representing attacks and as the input vector for machine learning methods such as multi-layer perceptron. Methods that use feature subsets relevant to specific attacks, therefore, have received considerable attention among many concerned researchers. In this respect, many methods based on the analysis of the correlations with attack classes such as the information gain [12], dependency ratio [13] and correlation [14] of individual features have been proposed. These methods eliminate features with lower ranks after ordering in terms of correlation measures. Although the correlation based methods guarantee efficiency, they cannot reflect the emergent effect of feature combinations, which is different from the naïve addition of individual features. The reason for almost all proposed feature selection methods dependent on the correlation analysis of individual features is that the number of possible feature subsets is too large to evaluate each feature subset through experimentation. For example, given a set of 41 features, the number of possible feature subsets is 241–1.

A method based on the meta heuristic algorithm was proposed to tackle the optimal feature selection problem by Kang et al. [15]. The proposed optimal feature selection algorithm is based on a local search algorithm to provide a feature subset for multi-layer perceptron. The authors showed that the feature subsets selected by the approach guarantees above 95% accuracy and the average size of feature subsets is 21, half of 41 features. However, they encountered two class problems in determining whether denial of service (DoS) attacks occur from the selected feature subset without specifing the kind of DoS attacks. In view of the fact that the contermeasure should be different according to the kind of attack, the functionality to discern the kind of attack is an important ability that IDSs require. Therefore, the research has limitations for applications to network-based IDSs in practice.

In this paper, we propose an optimal feature selection algorithm to characterize the six kinds of DoS attacks (neptune, teardrop, smurf, pod, back, land) defined in the KDD'99 data set and the NSL_KDD data set in addition to normal traffic. The proposed method is based on the simulated annealing algorithm. In order to evaluate the performance of selected feature subsets in terms of accuracy and efficiency using the proposed algorithm, three well-known machine learning techniques, experiments using multi-layer perceptron (MLP), naïve Bayes classifier and support vector machine (SVM), are carried out against the NSL_KDD data set. Subsequently, we compare the performance of our proposed method with that of the feature selection method based on the local search algorithm.

The paper is arranged as follows. In the second section, the composition and properties of the NSL_KDD data set is described. A feature selection algorithm based on the simulated annealing algorithm is proposed in Section 3. In Section 4, experiments using three machine learning methods are conducted to evaluate the performance of the selected feature subsets obtained by using the proposed feature selection method. A performance comparison is also carried out againt the NSL_KDD data set in terms of accuracy and efficiency. Lastly, we conclude and present future research in Section 5.

2 Material

2.1 NSL_KDD Data Set

We used the NSL_KDD data set [11] to evaluate the usability of the proposed feature selection algorithm. The KDD'99 data set [9], which has been widely used to evaluate IDSs, is composed of training data containing about 5 million samples and test data containing about 300,000 samples. Attacks in the data set are categorized into 4 classes (denial of service attack, user to root attack, remote to local attack, probing attack) including normal traffic. Additionally, 41 features (refer to Table 1) are classified into 3 groups: basic

Table 1   The 41 features of NSL_KDD data set

No	Feature Name	No	Feature Name
1	duration	22	is_guest_login
2	protocol_type	23	count
3	service	24	srv_count
4	flag	25	serror_rate
5	src_bytes	26	srv_serror_rate
6	dst_bytes	27	rerror_rate
7	land	28	srv_rerror_rate
8	wrong_fragment	29	same_srv_rate
9	urgent	30	diff_srv_rate
10	hot	31	srv_diff_host_rate
11	num_failed_logins	32	dst_host_count
12	logged_in	33	dst_host_srv_count
13	num_compromised	34	dst_host_same_srv_rate
14	root_shell	35	dst_host_diff_srv_rate
15	su_attempted	36	dst_host_same_src_port_rate
16	num_root	37	dst_host_srv_diff_host_rate
17	num_file_creations	38	dst_host_serror_rate'
18	num_shells	39	dst_host_srv_serror_rate
19	num_access_files	40	dst_host_rerror_rate
20	num_outbound_cmds	41	dst_host_srv_rerror_rate
21	is_host_login		

feature, contents feature and traffic feature. Among 41 features, features like duration, protocol_type, and service are classified as basic features and they can be usually extracted from TCP/IP connections. Features such as num_failed_logins, logged_in, num_compromised and su_attempted are part of the contents feature. Contents features are relevant to the attributes which help to detect a suspicious behaviour such as login failure. Lastly, the traffic feature is computed through observing the network connection using a time window of 2 seconds and it is classified into 2 categories: same host features and same service features. While serror_rate and rerror_rate are part of the same host feature, srv_error_rate and srv_rerror_rate are classifed under the same service feature. The 41 features detailed in the NSL_KDD data set are presented in Table 1.

The fact that the complete KDD'99 data set is too large makes it difficult to use the data set to compare the performance of proposed methods without artificial manipulation such as arbitrary selection of part of the data set according to the author's subjective decision. In addtion to the data size problem, the fact that the results of experiments could be biased towards the relatively abundant attack records has been revealed by many studies focusing on the KDD'99 data set itself.

To complement the disadvantages of the KDD'99 data set, M. Tavallaee et al. [10] proposed the NSL_KDD data set. While the NSL_KDD data set is basically a subset of KDD'99, it improves the KDD'99 data set as follows. Firstly, the NSL_KDD data set eliminates data redundancy such that it prevents the results of experiments from having a bias towards relatively redundant attack records. In addtion, the NSL_KDD data set increases the objectivity of the performance comparisons by adjusting the difficulty levels between attack classes. Lastly, the NSL_KDD data set consists of a reasonable amount of records so that objective comparisons among different detection methods are possible while avoiding the arbitrariness that occurs when randomly selected parts of the data set are used.

Because the goal of this paper is to propose a feature selection method for determining which kind of DoS attack occurs in a network among six kinds of attacks, normal records and six kinds of DoS attack records were extracted from the complete NSL_KDD data set. The selected part of the data set is composed of training data and test data, containing 113271 records and 15452 records, respectively. The composition of the prepared data set for experiments including normal instances is shown in Table 2. From the table, we can observe large differences in the number of records for DoS attacks in the NSL_KDD data set.

Table 2   The compostion of the data set

	Normal	Neptune	Teardrop	Smurf	Pod	Back	Land
Training data	67344	41214	892	2646	201	956	18
Test data	9711	4657	12	665	41	359	7

2.2 Data Preprocessing

NSL_KDD data set contains a range of feature types. Therefore, each feature needs to be normalized into a certain range of numeric values in order to be used as an input to the machine learning classifier. Data normalization was conducted following the method proposed in [2].

1) symbolic features like *protocol type* – integers from 0 to N–1, where N is the number of symbols, were assigned to each symbol and then each value was linearly scaled to the range of [0, 1].
2) numeric features with large integer value ranges like *src_bytes* and *dst_bytes* – logarithmic scaling with base 10 was applied to the features.
3) boolean features – the corresponding value of 0 or 1 was used without any modification.
4) all other numeric features: linearly scaled to the range of [0, 1].

After mapping to a certain range of numeric values, min-max normalization was applied to each scaled feature value. A feature value s is linearly transformed to a value in the range of [0, 1] using (1)

$$\frac{s - \min(f_i)}{\max(f_i) - \min(f_i)}, \tag{1}$$

where, $min(f_i)$ and $max(f_i)$ denotes the minimum value and the maximum value of the i-th feature over the training and test data set ($1 \leq i \leq 41$).

3 Optimal Feature Selection Algorithm

3.1 Optimal Feature Subset Selection Problem

Kang et al. [15] defined the feature selection problem for IDS as a combinatorial optimization problem. The number of possible feature combinations from the feature set with 41 features is 2^{41}–1. This number means that it is impossible to conduct performance evaluations for all feature combinations. The optimal feature subset selection problem is defined as follows:

Definition 1. Optimal feature subset selection problem

Given a feature set $f = \{f_1, f_2, f_3, \ldots, f_n\}$ and a cost function $C{:}f \rightarrow q$ ($0 \leq q$), find feature subset (s) f' such that the value of the cost function is minimized.

3.2 Feature Selection Approach based on Simulated Annealing Algorithm

We designed a novel feature selection algorithm for IDS. The proposed feature selection algorithm is based on simulated annealing, which is a widely used method in combinatorial optimization. The simulated annealing algorithm can be considered one of the search algorithms. However, while naïve local search algorithms use a greedy approach to find the optimal solution, simulated annealing is a probabilistic technique that enables us to leave the local optima to find better solutions. For this reason, simulated annealing is known to behave better than the naïve local search algorithm most of the time.

3.2.1 Solutions

A solution used in the feature selection algorithm is represented by a binary vector f with a length of 41 as (2). While the value of 1 is assigned to the selected feature, 0 is assigned to the unselected feature.

$$f = < f_1, f_2, f_3, \ldots, f_{41} >, where f_i \in \{0, 1\}, 0 \leq i \leq 41 \qquad (2)$$

Most search algorithms used to handle the optimization problem need an initial solution. The simulated annealing approach needs an initial solution as well. We randomly selected a feasible solution and used it as an initial solution.

In the meantime, the neighboring solutions for a given solution are defined as binary vectors with one bit different from the given solution. For example, the neighboring solutions for a solution which uses all 41 features are composed of 41 binary vectors with only one bit of value 0. $<0, 1, 1, 1, \ldots, 1>$ is one of them.

3.2.2 Cost function

One of the important factors on which the performance of optimization heuristic algorithm such as simulated annealing depends is the cost function for evaluating individual solutions. In other words, the performance of an algorithm is largely dependent on how the cost function is defined. The cost function used in this paper is similar to the one suggested in [15]. The basic idea of the cost function is to use the accuracy of clustering by using features represented by the given solution. How accurately the training data is partitioned into correspondig clusters when the clustering is conducted using only features presented by a solution serves as the cost for a given solution.

K-means clustering algorithm was adopted as a clustering algorithm for the cost function. The objective of the k-means clustering algorithm is to partition observation into k clusters such that the sum of the variation in the clusters is minimized. Because the paper dealt with classifying 6 kinds of DoS attacks and normal traffic, the feature selection problem belongs to the 7-class problem and the value of k is 7. Additionally, the cost φ for a given record x in the training data is computed by

$$\varphi(x) = \begin{cases} 1 & if\ p\,(x) = q(x) \\ 0 & otherwise \end{cases} \tag{4}$$

$\varphi(x)$ is set to 1 if the class $p(x)$ determined by the clustering algorithm is equal to the original class $q(x)$ that a record x belongs to, otherwise it is set to 0. The cost $C(f)$ for a given solution f is calculated over the training data set (with a size of N) using (5).

$$C(v) = \frac{1}{\sum_{i=1}^{N} \varphi(x_i)} \tag{5}$$

3.2.3 Other parameters

Simulated annealing adopts a cooling scheme to find optimal solutions avoiding local optima while searching the solution space. In general, the cooling scheme refers to a schedule for how to search. Parameters such as an initial temperature, a temperature reduction function and a termination condition have to be specified.

The initial temperature T has to be large enough to allow sufficient transitions to be accepted. A value of 100,000 was assigned as the initial value T, which is larger than the size of the training data set. The temperature reduction function is defined as a simple iterative function which is the product of T multiplied by a constant r.

$$T \leftarrow r \times T \tag{6}$$

where the value of r is set to 0.9. Lastly, the termination condition is that if the value of T is less than 0.001 then the algorithm stops where 0.001 was determined after several experiments.

3.2.4 Procedure of algorithm

The feature selection algorithm based on simulated annealing proceeds as follows.

Firstly, an initial solution is selected randomly and it is assumed to be the optimal solution. Subsequently, the cost of the initial solution is computed using the cost function described previously. While temperature T does not satisfy the termination condition, a neighboring solution of the current optimal solution is selected and its cost is calculated. If the cost of the newly selected neighboring solution is less than or equal to that of the current optimal solution, the current optimal solution is replaced with a newly selected neighbor solution. If the cost of the neighboring solution is greater than the current optimal solution, a random value q is chosen in the range of $(0, 1)$. In this case, the replacement of the optimal solution is permitted only if a random value q is less than $e^{-\frac{Cost(v_n)-Cost(v_b)}{T}}$. After temperature T is reduced following (6), these processes are continued until T satisfies the termination condition.

A pseudo-code for the feature selection algorithm based on simulated annealing is suggested as follows.

Algorithm: **Feature selection algorithm based on simulated annealing**
Input: Training data set
Output: Combination of features: v_b
1. $v_b \leftarrow$ Null;   // final solution
2. $T \leftarrow 100000$;
3. $r \leftarrow 0.9$;
4. Generate an initial solution, v_i;
5. $v_b \leftarrow v_i$;
6. Calculate the cost of initial solution, $Cost(v_b)$;.
7. while $(T > 0.001)$ do
8. begin:
9.   Randomly select a neighbor solution, v_n, of v_b which have one bit different from v_b;
10.   if $(Cost(v_b) = Cost(v_n))$:
11. $v_b \leftarrow v_n$;
13.   else:
14.   Generate a random number q uniformly in the range $(0, 1)$;
15.   if $(q < e^{-\frac{Cost(v_n)-Cost(v_b)}{T}})$
16. $v_b \leftarrow v_n$;
17. $T \leftarrow r \times T$
18. end // for while loop

---

4 Experiments and Results Analysis

We generated 20 feature subsets using the proposed feature selection algorithm for the performance evaluation of the selection algorithm.

4.1 Machine Learning Methods

In order to evalute the selected feature subsets, three representative supervised machine learning methods, MLP, Bayes classifier and SVM, were used. A brief introduction about machine learning methods including the used parameters are presented in this sub section.

4.1.1 Multi-layer perceptron

Multi-layer perceptron, also known as an artificial neural network, is a supervised machine learning method that learns and recognizes objects by imitating the information process of the human brain. A 3-layer format containing input, hidden and output layers was adopted as the basic structure of MLP. The number of nodes at the input layer is equal to the size of the given feature subset and the number of nodes at the output layer is set to 7. The number of nodes at the hidden layer was set up with a number which showed the best performance through many experiments. The sigmoid function was used as a neuron function and MLP was learned using the backpropagation algorithm with various learning rates and momentums.

$$S(v) = \frac{1}{1 + e^{-v}} \tag{7}$$

The learning algorithm terminated when the improvement with respect to error rate was less than 0.1% between two contiguous epochs over the entire training data set.

4.1.2 Naïve Bayes classifier

The naïve Bayes classifier [16] is a supervised machine learning classifier based on the following Bayes theorem.

$$p(\omega_i|x) = \frac{p(x|\omega_i)p(\omega_i)}{p(x)} \tag{8}$$

where w_i indicates the ith class and x denotes a given feature vector. In the naïve Bayes classifier, the class of the given feature is determined by the class with the greatest posterior probability $p(x|w_i)$. We assumed that the distribution of

$p(x|w_i)$ follows the Gaussian distribution and the maximum likelihood method was used to determine the parameters of the distribution.

4.1.3 Support vector machine

Support vector machine (SVM) is one of the supervised machine learning methods suggested by Vapnik [17]. Because it considers the margin between support vectors when finding the decision hyperplane, it is known to have generality compared with other machine learning methods.

We used a polynomial kernel function $K(x, y) = (x \cdot y + 1)^p$ for the non-linear support vector machine. Lagrange multipliers were obtained by a sequential minimal optimization (SMO) algorithm [18]. Originally, SVM was a 2-class classifier. The 2-class SVM for solving multi-class problems that deal with more than two classes had to be expanded. The pair-wise classification method that adopts voting to identify the given feature was used.

4.2 Performance Comparisons

In order to evelute the performance of the feature subset computed by the proposed selection algorithm, the accuracy, which is generally evaluated in studies on IDS, was measured. The accuracy is the proportion of correct results in determinations by the machine against the test data set. In additon to the accuracy, the time taken to train and test and the size of the feature subset were measured as well. These measures are important evaluation factors for real time IDS/IPSs.

Table 3 presents the average accuracy and standard deviation when three machine learning methods were applied to 20 feature subsets obtained by two feature selection algorithms, one based on the local search algorithm and the other based on simulated annealing. In the same table, the accuracy of all 41

Table 3 The results of experiments which show accuracy, time taken for training, time taken for testing and feature size

	Accuracy(%)		
	Multi-Layer Perceptron	Naïve Bayes Classifier	Support Vector Machine
Local search algorithm	96.77 (±1.52)	81.30 (±19.12)	97.15 (±2.13)
Simulated annealing	96.83 (±0.90)	86.64 (±14.40)	97.48 (±1.38)
All features	96.98	98.33	99.24

features when they were used in the three machine learing methods is also shown.

The accuracy of the feature subsets obtained by the proposed feature selection algorithm was 96.83% for MLP, 96.64% for the Bayes classifier and 97.48% for SVM. Although these accuracy values are lower than those obtained when all features were used, they are slightly higer than those obtained with the local search based feature selection algorithm. In the case of the naïve Bayes classifier, while the simulated annealing algorithm showed 5% higher accuracy than the local search algorithm, it was quite a bit lower than the accuracy when all 41 features were used. However, the best accuracy values achieved by the proposed algorithm for the three machine learning methods among 20 feature subsets were 98.74%, 98.69% and 99.24% respectively, which were higher than or equal to those achieved when 41 features were used. This means that if we seek many solutions with the proposed algorithm and select the solutions with high performance, we can achieve accuracy as high as that obtained when 41 featues are used for machine learning techniques regardless of the types. Table 4 shows the feature subsets which achieved the best accuracy values among the 20 feature subsets used for three machine learning techniques with their accuracy values.

Table 5 shows the average lengths of feature subsets obtained by two feature selection algorithms using MLP. It also shows the time required for training and the time for testing. The average length of feature subsets produced by the proposed feature selection algorithm is 18 ($\pm$3.74) and is lower than that of feature subsets obtained by the feature selection algorithm based on a local search algorithm. As we can expect, the time taken to train and test the proposed method is shorter than that of the local search based algorithm. The length of the feature vector and especially the time taken to determine whether an attack occurs or not, even though it is trivial, is as important as the accuracy in realtime IDS/IPSs. The best solution in terms of

Table 4   A feature set with accuracy of 98.8%

	Feature Compositon	Accuracy
MLP	0, 0, 0, 0, 1, 0, 0, 1, 0, 1, 1, 1, 0, 1, 0, 0, 0, 1, 0, 0, 0, 0, 0, 1, 0, 1, 0, 1, 1, 0, 0, 0, 1, 0, 0, 0, 1, 1, 0, 0	98.74
Bayes	1, 1, 0, 0, 1, 1, 0, 0, 0, 1, 1, 1, 0, 1, 0, 1, 0, 0, 0, 1, 0, 1, 0, 0, 0, 1, 1, 0, 0, 1, 0, 1, 1, 0, 0, 0, 0, 0, 0, 0	98.69
SVM	0, 1, 0, 1, 1, 0, 0, 1, 0, 1, 1, 1, 0, 0, 0, 0, 1, 1, 0, 1, 1, 0, 1, 0, 1, 0, 0, 1, 1, 0, 0, 0, 0, 0, 0, 1, 1	99.24

	Length of Feature Vector	Time for Training	Time for Testing
Local search algorithm	19.05 (±3.47)	315.99 (±49.31)	0.71 (±0.12)
Simulated annealing	18.00 (±3.74)	296.79 (±46.19)	0.64 (±0.12)
All features	41	799.65	1.48

Table 5 The average lengths of feature subsets

the time had a length of 14 and the time taken to train and test was 233.11 sec and 0.49 sec, repectively.

5 Conclusion

In this paper, we proposed an optimal feature selection algorithm for detecting six kinds of denial of service attacks against the NSL_KDD data set. The feature selection problem was defined as a combinatorial optimization problem. The proposed algorithm is based on the simulated annealing algorithm.

In order to evaluate the accuracy and efficiency of selected feature subsets obtained by the proposed feature selection algorithm, MLP, Bayes classifier and SVM were used against the NSL_KDD data set. A comparison between our proposed algorithm and other feature selection algorithms was conducted including a comparison with the results obtained when all 41 feature sets were used. From the experiment results, we confirmed that the feature subsets selected by the proposed algorithm have a higher accuracy and detection rate. In addition, the average length of the feature subsets obtained by the proposed algorithm was 18 and the algorithm was more efficient in both learning and identifying time. This indicates that the proposed feature selection algorithm is suitable for realtime IDS/IPS.

References

[1] Paliwal, S., and Gupta R. (2012). Denial-of-service, probing & remote to user (R2L) attack detection using genetic algorithm. *Int. J. Comput. Appl.* 60, 57–62.

[2] Sabhnani, M., and Serpen, G. (2003). Application of machine learning algorithms to KDD intrusion detection dataset within misuse detection context. *Proc. Int. Conf. Mach. Learn. Model Technol. Appl.* 209–215.

[3] Bankovic, Z., Stepanovic, D., Bojanic, S., and Nieto-Taladriz, O. (2007). Improving network security using genetic algorithm approach. *Comput. Electr. Eng.* 33, 438–451.

[4] Azad, C., and Jha, V. K. (2013). Data mining in intrusion detection: a comparative study of methods, types and data sets. *Int. J. Inf. Technol. Comput. Sci.* 5, 75–90.

[5] Balajinath, B., and Raghavan, S. V. (2001). Intrusion detection through learning behavior model. *Comput. Commun.* 24, 1202–1212.

[6] Tsai, C. F., Hsu, Y. F., Lin, C. Y., and Lin, W. Y. (2009). Intrusion detection by machine learning. *Rev. Expert Syst. Appl.* 36, 11994–12000.

[7] Wu, S. X., and Banzhaf, W. (2010). The use of computational intelligence in intrusion detection system. *Rev. Appl. Soft Comput.* 10, 1–35.

[8] Kolias, C., Kambourakis, G., and Maragoudakis, M. (2011). Swarm intelligence in intrusion detection: a survey. *Comput. Secur.* 30, 625–642.

[9] KDD Cup. (1999). Available at: http://kdd.ics.uci.edu/databases/kddcup9 9/kddcup99.html.

[10] Tavallaee, M., Bagheri, E., Lu, W., and Ghorbani, A. A. (1999). "A Detailed Analysis of the KDD CUP 99 Data Set," in CISDA'09 Proceedings of the Second IEEE international conference on Computational intelligence for security and defense applications, Ottawa, ON (NJ, USA: IEEE Press Piscataway), 53–58.

[11] NSL_KDD data set. Avalilable at: http://nsl.cs.unb.ca/NSL-KDD/

[12] Kayacik, H. G., Zincir-Heywood, A. N., and Heywood, M. I. (2005). "Selecting features for intrusion detection: a feature relevance analysis on kdd 99 intrusion detection datasets," in *Thrid Annual Conference on Privacy, Security and Trust*.

[13] Olusola, A. A., Oladele, A. S., and Abosede, D. O. (2010). "Analysis of KDD'99 intrusion detection dataset for selection of relevance features," in *Proceedings of the World Congress on Engineering and Computer Science*, Vol. 1.

[14] Parazad, S., Saboori, E., and Allahyar, A. (2012). "Fast feature reduction in intrusion detection datasets," in *MIPRO, Proceedings of the 35th International Convention* pp.1023–1029.

[15] Kang, S.-H., and Kim, K. J. (2015). A feature selection approach to find optimal feature subsets for the network intrusion detection system. *Cluster Comput.* doi: 10.1007/s10586-015-0527-8

[16] John, G. H., and Langley, P. (1995). "Estimating continuous distributions in Bayesian classifiers," in *UAI'95 Proceedings of the Eleventh conference on Uncertainty in artificial intelligence.*

[17] Burges, C. J. C. (1998). A tutorial on support vector machines for pattern recognition. *Data Min. Knowl. Discov.* 2, 121–167.

[18] Platt, J. C. (1998). *Sequential minimal optimization: a fast algorithm for training support vector machines.* Technical Report MSR-TR-98-14.

Biographies

I.-S. Jeong received her M.S. and Ph.D. degree in computer science from the Chonnam National University, Gwangju, Korea, in 2006 and 2011. During 2011–2015, she was a postdoctoral researcher in genomics division at Rural Development Administration, Korea. Her research interests include machine learning, data mining, algorithms in bioinformatics, and sensor networks.

H.-K. Kim is a professor of Information Security Department in the Dongshin University, Naju, Korea. He received his M.S. and Ph.D. in Computer Science from Chonnam National University, Gwangju, Korea, in 1986 and 1996, respectively. His research interests include information security, spatial data structure, and graphics.

T.-H. Kim received her M.S. and Ph.D. in Computer Science from Chonnam National University, Gwangju, Korea, in 1991 and 1999, respectively. During 1993–1997, she was a part time lecturer at Dongshin University, Naju, Korea. She joined Dongshin University, Naju, Korea, in 1998, where she works as an associate professor. Her research interests include information security, security programming, and database security.

D. H. Lee received the B.S. degree in Computer Science from Kyonggi University, Korea. He received M.S. and Ph.D degree in Information Security from Kyonggi University, Korea. and Research Scholar of University of Colorado Denver, USA, in 2011 and 2012. He is currently a assistant Professor in Information Security, Dongshin University, Korea.

His research areas include Information Security and Convergence security.

K. J. Kim is a professor of Information Security Department in the Kyonggi University, Korea. He received his Ph.D and MS in Industrial Engineering from Colorado State University in 1994. His B.S in Mathematics from the

University of Kansas. He is Executive General Chair of the Institute of Creative and Advanced Technology, Science, and Engineering. His research interests include cloud computing, wireless and mobile computing, digital forensics, video surveillance, and information security. He is a senior member of IEEE.

S.-H. Kang received his M.S. and Ph.D. in Computer Science from Chonnam National University, Gwangju, Korea, in 2003 and 2009, respectively. During 2010–2013, he was a researcher in the National Institute for Mathematical Science, Daejeon, Korea. He joined Dongshin University, Naju, Korea, in 2013, where he works as an assistant professor. His research interests include information security, wireless sensor networks, and algorithm.

Improving Intrusion Detection on Snort Rules for Botnet Detection

Saiyan Saiyod[1,*], Youksamay Chanthakoummane[1],
Nunnapus Benjamas[2], Nattawat Khamphakdee[2]
and Jirayus Chaichawananit[1]

[1]*Hardware-Human Interface and Communications (H2I-Comm) Laboratory,
Department of Computer Science, Faculty of Science, Khon Kaen University,
Muang, Khon Kaen, Thailand*
[2]*Advanced Smart Computing (ASC) Laboratory, Department of Computer Science,
Faculty of Science, Khon Kaen University, Muang, Khon Kaen, Thailand*
*E-mail: {saiyan; nunnapus}@kku.ac.th; {youksamay_c; jirayus.chaichawananit;
k.nattawat}@kkumail.com*
*Corresponding Author

Received 10 April 2016; Accepted 2 May 2016;
Publication 29 May 2016

Abstract

The Botnets has become a serious problem in network security. An organization should find the solutions to protect the data and network system to reduce the risk of the Botnets. The Snort Intrusion Detection
System (Snort-IDS) is the popular usage software protection of the network security in the world. The Snort-IDS utilizes the rules to match
the data packets traffic. There are some existing rules which can detect
Botnets. This paper, improve the Snort-IDS rules for Botnets detection
and we analyze Botnets behaviors in three rules packet such as Botnets_
attack_1.rules, Botnets_attack_2.rules, and Botnets_attack_3.rules. Moreover, we utilize the MCFP dataset, which includes five files such as
CTU-Malware-Capture-Botnet-42, CTU-Malware-Capture-Botnet-43, CTU-
Malware-Capture-Botnet-47, CTU-Malware-Capture-Botnet-49, and CTU-
Malware-Capture-Botnet-50 with three rule files of the Snort-IDS rules. The
paper has particularly focused on three rule files for performance evaluation

Convergence Security, Vol. 1, 19–40.
doi: 10.13052/jcs2445-9992.2016.002

of efficiency of detection and the performance evaluation of fallibility for Botnets Detection. The performance of each rule is evaluated by detecting each packet. The experimental results shown that, the case of Botnets_attack_1.rules file can maximally detect Botnets detection for 809075 alerts, the efficiency of detection and fallibility for Botnets detection are 94.81% and 5.17%, respectively. Moreover, in the case of Botnets_attack_2.rules file, it can detect Botnets up to 836191 alerts, having efficiency of detection and fallibility for Botnets detection are 97.81% and 2.90%, respectively. The last case Botnets_attack_3.rules file can detect Botnets 822711 alerts, it can 93.72% of efficiency of detection and the value of fallibility is 6.27%. The Botnets_ attack_2.rules file is most proficient rule for Botnets detection, because it has a high efficiency of detection for detection and a less of fallibility.

Keywords: Botnets detection, intrusion detection system, MCFP datasets, Snort-IDS.

1 Introduction

Currently, a transaction has mainly exchanged the information through the Internet, such as the purchase, sale and product exchange. However, the data security has become the most significant tool for personal computers, home, school, business, and offices. The events have been serious impact for business operations. If information is stolen, it will damage and the system needs to stop immediately. The current threat of attacks of security forces have increased. Moreover, the threats of the Botnets in different ways have been developed. Therefore, the implementations of business organizations via the Internet have increased the network security to protect the information efficiently. Thus, the attackers produce the information deviation and make network system damage [1].

Botnets or Bot is the robot software that is usually installed on the client computer and run the command preset automatically. For example, IRC (Internet Relay Chat) is Bot of chat rooms and Bot of games online. The Botnets is controlled remotely by hacker. Bot is a small type of computer virus, so users cannot notice that the Bot in computer. These computers are often controlled by hackers via the Internet and send orders through the system on the chat room or IRC [2]. The Botnets are malicious attacking and infecting, sending SPAM, DDoS, Blacklist, Neris, etc. Furthermore, the Bot is remotely managed and synchronization. The synchronization is not only between different Bot, but it is also within the same Bot. For example, the

Bot module sends the SPAM and the module maintains the C&C channel and stops sending packets at the same time when the Bot receives the order of not sending any more mails [3].

Today, an intrusion detection system (IDS) is able to maintain various kinds of techniques in network security. The main task of IDS is to monitor the network traffic data on the network system and analyze it. If any suspicious files alert is when a malicious attacks the system. Besides, the IDS can be presence of signature-base and anomaly-base. Another limit is easy to maintain the network security based on accuracy [4].

This paper, we propose the improvement of the Snort-IDS rules for Botnets detection and utilizing MCFP datasets with rules techniques. The rule techniques analyze the network traffic data pattern, which Snort-IDS are generated according to the network's traffic data behavior.

All the information has been clarified as following details. The Section 2 present related work. The Section 3 presents the background. The Section 4 present system architecture. The Section 5 present evaluation performance. The Section 6 present conclusion and future work.

2 Related Work

Botnets particularly, pose a significant threat to the security on the Internet. As a result, there are many interests in the research community to improve sufficient solutions. In paper [5], the IDS-base and multi-phase IRC-Bot and Botnets behavior detection model are based on C&C reopens message and malicious behaviors of the IRC-Bot inside the network environment. Moreover, a detection method for detecting Botnets is based on behavior features. It's capable of detecting in both known and unknown Botnets and it is less updates in the research fields of Botnet detection [6]. The intrusion detection system was implemented with Snort and configured with WinPcap, within Windows-based environment. It was possible to configure it as a firewall on the Windows. The Snort-IDS rules, however, were not improved [7].

The intrusion detection on the network Botnets attacks were improved through the utilization of the MCFP (Malware Capture Facility project) datasets [3]. The authors proposes three new Botnets detection method and the new model of Botnets behavior, which are based on a deep understanding of the Botnets behavior in the network such as the SimDetec, the BClus method, and the CCDetect. These algorithms can access a better datasets to start showing the particular result. The shift of the detection techniques for behavior base models has proved to be a better approach to the analyze Botnets pattern. However, the current knowledge of the Botnets detection and the pattern does not have an obvious analysis.

Additional, utilizing of Snort-IDS monitor Web content in a certain time by identifying the abnormal behavior patterns within a campus network security monitoring system [8]. The performance evaluation results generate alert analysis of the Snort-IDS in high-speed network which means that Snort has detected 12 signatures among which detection ICMP PING attacks [9]. In model Snort-IDS structure analyzes the pattern synchronize the protocol or improve speed and accuracy of intrusion detection system in campus network. ACID stand for Analysis Center Intrusion Detection which chosen to shows alarm information [10]. Nevertheless, all research that mention can detect the Botnets, but those research cannot improve the Snort-IDS rules.

3 The Background

We discuss the background of Snort-IDS, the MCFP datasets, and the background of Botnet.

3.1 The Background of Snort-IDS

Snort is useful software for security network. In 1998 Matin Roesch developed the Snort Intrusion Detection System and Intrusion Prevention System (Snort-IDS/IPS) by using C language as an open-source software and lightweight software application. Snort can be installed on numerous platforms of operating systems such as Windows, Linux, etc. Snort has a real time alerting the traffic data network and analyzes capability. The alert will be sent to syslog or a separated 'alert' files, or to popup windows. Snort is logically divided into multiple components. These components working step by step process of detecting particular attacks and to generate output in required format from the detection system. The components of Snort are packet decoder, preprocessors, detection engine, logging and alerting system, and output modules [11]. Snort utilize with rule to alert the network traffic data. The Snort-IDS rules have two logical parts such as the rule header and the rule option [12] as shown in Figure 1.

| Rule Header | Rule Option |

Figure 1   Structure of Snort rules.

The rule header. The rule header describes attributes of a packet and to command the Snort what to do when it founding the packet that matches the rule as shown in Figure 2.

The rule options. The rule options will follow the rule header and can alert message, information on which part of packet.

The example rule as shown in Figure 3. The network traffic data of TCP protocol for example source address is any, source port is 21, destination address is 10.199.12.8, destination port is any, an generate that outputs the message "TCP Packet is detected" with signature id:1000010.

3.2 The Background of MCFP Datasets

The Malware Capture Facility Project (MCFP) datasets [3, 13]. The MCFP were capture in the CTU University in Czech Republic. The datasets have large size, so they are stored in the server in the university. The goals of datasets were to have a large capture of Botnets traffic mixed with normal traffic and background traffic. The totals of MCFP datasets are 13, but for this research we use 5 datasets.

- CTU-Malware-Capture-Botnet-42 is dataset corresponds to an IRC-based Botnets to send spam for almost six and a half hours and the completed Pcap size is 56 MB, total of Botnets in datasets are 323154.

Action	Protocol	Source Address	Source Port	Direction	Destination Address	Destination Port

Figure 2   Structure of Snort rule header.

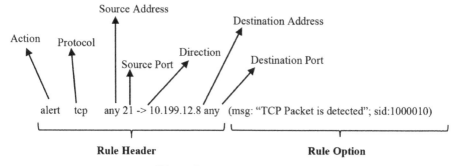

Figure 3   Snort rule example.

- CTU-Malware-Capture-Botnet-43 corresponds to an IRC-based Botnets to send spam for 4.21 hours and complete Pcap size as 30 MB, total of Botnets in dataset are 176064.
- CTU-Malware-Capture-Botnet-47, the Botnets in this scenario scanned SMTP (Simple Mail Transfer Protocol) servers for two hours and connected to several RDP (Remote Desktop Protocol) services. However, it does not send any SPAM and attack. The C&C server uses proprietary protocols that connects every 33 seconds and send an average of 5,500 bytes on each connection. It has Pcap size as 5.0 MB and the total of Botnets in dataset are 24764.
- CTU-Malware-Capture-Botnet-49, the Botnets contact a lot of different C&C hosts with Chinese-based IP addresses and the Blacklist DNS. It receives the large amounts of encrypted data and Pcap size as 20 MB and the total of Botnets in dataset are 85735.
- CTU-Malware-Capture-Botnet-50, ten hosts were infected using the same Neris botnet as in scenario 1 and 2. For five hours more than 600 SPAM mails can be successfully sent. It has completes Pcap size as 1.0 GB and the total of Botnets I dataset are 259949.

3.3 The Background of Botnets

Botnets are the technological backbone supporting myriad of attack, including identity stealing, organizational spying, DDoS, SPAM, and government-sponsored attacks. Botnet is a network interface machine which aims to disseminate malicious code over the Internet without user intrusiveness. There are many types of Botnets that shows some serious attacks follow as;

- The Command and Control channel (C&C channels) is the architecture of the C&C Botnets which is a serious attack when the attack would not reveal the name of the attacker. In addition, the infected machines (Bots) receive instructions form C&C and respond it depend on those in striations. The instructions/commands rang from initiating a worm or spam attack over the Internet to disrupt a legitimate user request [14].
- IRC Bot. Many of these IRC Bot is passed by undetected until they become a significant problem. There are several reasons for this. For example they do not follow the same pattern of contagion, some state full firewall both hardware and the application might not alert this traffic, until it is initiated at the client side once the compromise has taken the place. IRC has several forms including Neris, Rbot, SPAM, Virut, NSIS, Menti, Sogou Murlo etc [14].

4 System Architecture

The system architecture, we proposed system consist of the following details; improving of Snort-IDS rules procedure, analysis Botnets, improved Snort-IDS rules, and proposed Snort-IDS rules.

4.1 Improving of Snort-IDS Rules Procedure

The Snort rules evaluation procedure, the MCFP datasets are utilized to test and evaluate detection performance. The datasets were recorded with various amounts of Botnets. In this paper, we utilize .Pcap file type, which contains the traffic data such as source address and destination address, source port and destination port, time to live (TTL), Ack, flags and so on. All attributes that mention are very important parameters for analyzing the attacking type and improving Snort-IDS rules. It would need to increase the perfection of the detection rules and decrease false alert [14] shown as Figure 4.

The Snort-IDS sensor was installed by using the Snort version 2.9.2.2, which runs on the Linux CentOS 64 bit version 6.4 operating system. The

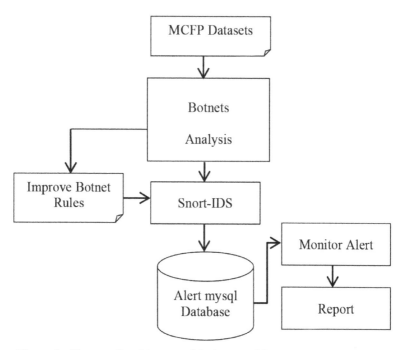

Figure 4 The overall architecture of the improved Snort-IDS rules procedure.

architecture configurations consist of Mysql database, Intel(R) Core(TM)2 Quad 2.66GHz CPU, 4GB DDR-RAM, 250GB HDD and Marvell 88E8071 GB Ethernet 10/100/1000 BaseT [15].

4.2 Botnets Analysis

Botnets analysis is a significant part of the system because it acts as a means to analyze and convert the traffic data of a network. Tokens Matching Algorithm is proposed to analyze the data format to be used for creating the new rules.

- Creating Tokens will reform the structure of the interested in traffic data to be the Tokens form.

- Comparing with Data Dictionary is to compare the created Tokens with Data Dictionary. If the created Tokens is matched by the Data Dictionary, the new will be created by considering the corresponding contents.

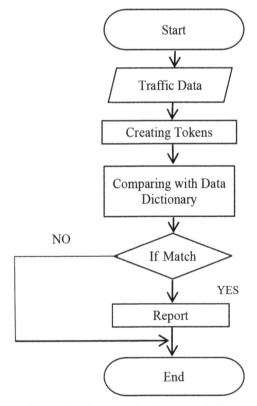

Figure 5   Flow chart for Botnets analysis.

Table 1 The set of letters, number, and symbols for creating the tokens

Tokens	Symbols
\<c\>	a b c d e f g h t i j k l m n o p q r s t u v w x y z
\<C\>	A B C D E F G H I J K L M N O P Q R S T U V W X Y Z
\<N\>	0 1 2 3 4 5 6 7 8 9
\<a\>	. , - :

Note: Table 1. shows the set of letters, numbers, and symbols that can be compared with contents in traffic data.

tokenizer (char) return (token)
 if is_small (char)
  return 'c'
 else if is_big (char)
  return 'C'
 else if is_num (char)
  return 'N'
 else
  return 'a'
 end
end

Figure 6 Function of creating tokens.

4.3 Improved Snort-IDS Rules

We compare the analytical information of data packets between each packet connection to dataset with attacking event example type of Botnet in dataset, time for capture, and number of Botnets capture in datasets, etc. [13]

4.4 Proposed Snort-IDS Rules

We explain more detail of Snort-IDS rules which are utilized for Botnet detection.

5 Evaluation of the Performance

In this section, we propose the experimental evaluation of the Snort-IDS rules to compare the detection performance. Evaluation of the performance consist of two procedure are the evaluation of the Snort-IDS rules procedure and detection accuracy comparison of the Snort-IDS rules.

No.	Name	Contents and Tokens								
1	IRC_SPAM	Con =	■	■	1	2	1	2	1	2
		Tok =	c	c	■	■	■	■	■	■
2	C&C_Botnets	Con =	■	■	■	.	m	s	-	p
		Tok =	c	c	c	a	■	■	a	■
3	Possible_NBSS Neris.exe	Con =	■	■	■	■	■	■	.	v	e
		Tok =	c	c	c	c	c	a	■	■
4	Win32.Domsingx. A contact to C&C server attempt	Con =	■	■	■	■	■	■	-	
		Tok =	c	c	c	c	c	c	a	C
5	TT-bot botnet contact to C&C server attempt	Con =	■	■	-	■	■	■	■	■
		Tok =	c	c	a	C	c	c	c	c
6	Paleo bot DNS request for C&C attempt	Con =	e	n	g	t	h	:		■
		Tok =	■	■	■	■	■	a	a	N
7	Botnets attack Chiness_C&C_ho sts on http	Con =	■	■	■	-	■	■	■	■
		Tok =	c	c	c	a	C	c	c	c
8	Neris_IRC sent to HTTP Attacks	Con =	.	.	.	■	■	■	■	■
		Tok =	a	a	a	N	N	N	N	N
9	BLACKLIST DNS request for known malware domain www...	Con =	■	■	■	■	■	■	■	.
		Tok =	c	c	c	c	c	c	c	a
10	Trojan.Win32.Shy lock.A contact to C&C server attempt..	Con =	■	■	■		1	2	1	2
		Tok =	c	c	c	c	a	a	C	c

■ Everything is possible or it is anything

☐ Like the contents

Table 2   Data dictionary

lookup_datadic (content, token) return (botnet_name_list)
  for i = 1 → botnet_number
       if bn_c [i] == content & bn_t [i] == token
then
                bonet_name_list += bn_n [i]
           end
        end
     end
return botnet_name_list

Figure 7 Comparing with data dictionary function.

5.1 The Evaluation of the Sort-IDS Rules Procedure

MCFP dataset that has been tested to evaluate performance the intrusion detecting of S. García [3], which has been stored in various file formats. In the datasets will contain Background files, Normal files and Botnet files. In this paper, we choose .Pcap which files have capacity the number of Botnet attacks. The Snort system is installed by utilizing Snort version 2.9.2.2, which runs on the Linux CentOS system [14]. The experiment results of the Snort rules need to modified *snort.conf* file to make it conform to the dataset. We also utilize database mysql for storage the alert data.

In Figure 8, shows the Snort testing procedure. We take command of *Snort –N –r botnet-capture-20110810-neris.pcap –c /etc/snort/snort.conf* to direct Snort-IDS to process dataset file [14]. We explain some command lines, for *–N* is option to make sure that the Snort does not log each packet to the terminal, for *–r* is option which .Pcap must be loaded, and the last one for *–c* option to specify where the config file is located. When the Snort-IDS detected traffic packet which match or synchronize to the Snort rule. The system will generate alert and to bring the recode alert into the database.

Performance evaluation of the Botnets_attack_1.rules, Botnets_attack_2. rules, and Botnets_attack_3.rules files can detect many Botnets such as Neris-IRC, IRC to send spam, Blacklist malware, generic IRC Botnets connection attempt, and malicious URI etc. The numbers of alert are shown in Table 3, Table 4 and Table 5.

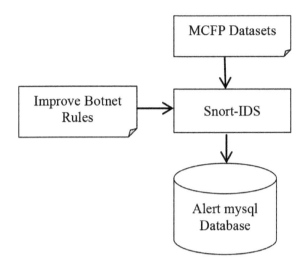

Figure 8   Evaluation of the Snort-IDS rules procedure.

Table 3   Number of Snort-IDS rules

No.	Type of Botnets Rules Files	Number of Rules
1	Botnets_attack_1.rules	389
2	Botnets_attack_2.rules	286
3	Botnets_attack_3.rules	273

Note: Improving the Snort-IDS rules already, all rules are saving or recorded in text file format. Nevertheless, there are numerous ways of the Botnets detection. Thus, we categorize the Botnet detection into 3 groups as shown in Table 3.

Table 4   Some example proposed Snort-IDS rules

No.	Type of Botnets Rules
1	Rule 1: alert tcp any any -> any any (msg:"Possible_NBSS Neris.exe SENT TO BROWSER"; flow:to_server,established; content:"ACACACAC"; ttl:128; sid:2000341; rev:1) Rules 2: alert tcp any any -> any any (msg:"Neris_IRC sent to HTTP"; flow:to_server,established; content:"7&_salt="; ttl:128; sid:2000366; rev:1)
2	Rules 1: alert tcp any any -> any any (msg:"TROJAN Possible IRCBot.DDOSCommon Commands"; flow:to_server,established; content:"AAC6F603"; ttl:128; sid:1000255; rev:7) Rules 2: alert udp any any -> any any (msg:"IRC-bot sent spam BROWSER"; flow:to_server,established; content:"FENEBEOC"; ttl:128; sid:1000199; rev:1)
3	Rules 1: alert tcp any any -> any any (msg:"Win32.Domsingx.A contact to C&C server_attempt"; flow:to_server,established; content:"ent-Type"; ttl:128; sid:4000309; rev:2)

Rules 2: alert tcp any any -> any any (msg:"BLACKLIST DNS request for known malware/domain/sxzyong.com";flow:to_server,established;content:"mail.com"; fast_pattern:only;/metadata:impact_flagred,service_dns;reference:url,www.virustotal. com/filescan/report.html?id=a7f97ed5c064b038279dbd02554c7e555d97f67b601b9 4bfc556a50a41dae137-1304614426; classtype:trojan-activity; ttl:128; sid:3000230; rev:2)

Note: Some of examples the Snort-IDS Rules are shown in Table 4. In No. 1, we improved which called "Neris.exe attacks". In this rule allows the Snort-IDS to attack in the Internet network connection. The Snort-IDS will to alert when the attacker scan the computer on the web, which open to any port. In Rules 2, we improve "Neris-IRC" which rule allow to the internet connection when the attacker send to website. Some of example No. 2, in Rules 1, we improve which call "TROJAN-IRC" which rules allow to Snort-IDS rule when the attacker to send command on the computer. Rules 2, we improved which called "IRCBot". This is rule allow the Snort-IDS network. The Snort-IDs will to alert when the attacker send spam to browser by UTP protocol packets which open any port and it have time to live (TTL) 128.

Some of example No. 3, in Rule 1, we improve alert by TCP protocol packets when the attacker to send "Win32" to C&C server, which open any port and any source address and destination port. In Rule 2, we improve the attacking detection rules for "Botnet BLACKLIST Malware" which is newly added to the dataset. The attacker utilize BLACKLIST DNS request for known malware domain on this website: sxzyong.com.

Table 5   The performance evaluation Botnets_attack_1.rules file

No.	Rules Type	Botnets	Number of Alert	Total Detected Number
1	CTU-Malware-Capture-Botnet-42	IRC_SPAM	192202	
		C&C_Botnet	2279	
		Possible_NBSS Neris.exe	9324	
		Neris_IRC sent to HTTP Attacks	92162	
		BLACKLIST DNS	747	296714
2	CTU-Malware-Capture-Botnet-43	Possible_NBSS Neris.exe	112679	
		C&C_Botnet	1403	
		IRC_SPAM	5817	
		Neris_IRC sent to HTTP	51402	
		BLACKLIST DNS	72	171373
3	CTU-Malware-Capture-Botnet-47	BLACKLIST DNS	23430	
		Spyeye bot contact to C&C server	44	
		possible_SPAM	310	
		Neris	98	
		C&C_Botnet	226	24108
4	CTU-Malware-Capture-Botnet-49	C&C_Botnet	79006	
		Virut DNS request for C&C attempt	4034	
		Neris_IRC sent to HTTP	188	83235

(Continued)

Table 5   Continued

No.	Rules Type	Botnets	Number of Alert	Total Detected Number
5	CTU-Malware- Capture-Botnet-50	IRC_SPAM Neris C&C_Botnet	102081 131273 291	 233645
Total				809075

Note: The performance evaluation shows in Table 5 are very efficiency for Botnets detection. The Botnets_attack_1.rules file consist the event of Botnet types which are IRC-bot, BLACKLIST, SPAM, and NBSS Neris etc. The number of detection may not be enough, however the rule file can help administrators know how attacks from Botnets.

5.2 Detection Accuracy Comparison of the Snort-IDS Rules

The procedure of comparing the accuracy of Snort-IDS rule for Botnets detection. The information alert from database will be compared with events of actual invasion. The information is available in the Detection Scoring Truth, Which we used number of Botnet in the MCFP datasets and the Botnets have been alert in our three rule files (Botnets_attack_1.rules, Botnets_attack_2.rules, Botnets_attack_3.rules). The information are compared with experimental information and having the most accurate. We does not include, resolve in this section, more detail shows as Figure 9.

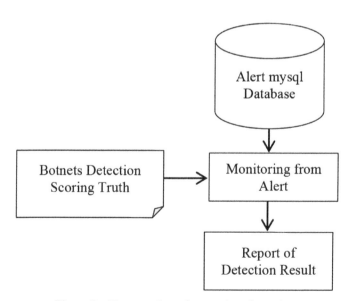

Figure 9   The procedure of comparison from alert.

Table 6 The performance evaluation of Botnets_attack_2.rules file

No.	Rules Type	Botnets	Number of Alert	Total Detected Number
1	CTU-Malware-Capture-Botnet-42	Ozdok botnet communication C&C server	170	
		IRC_Bot	84	
		BLACKLIST DNS	29	
		Possible IRC-bot-Sent Spam	145427	
		Neris	165674	311384
2	CTU-Malware-Capture-Botnet-43	Neris Botnet	74710	
		IRC-bot sent spam BROWSER	94693	
		Botnets attack Chinese_C&C_hosts on http	123	169526
3	CTU-Malware-Capture-Botnet-47	BLACKLIST USER-AGENT	1522	
		Spyeye bot contact to C&C server attempt	9266	
		C&C_Botnets	13589	
		IRC sent_spam	145	24522
4	CTU-Malware-Capture-Botnet-49	Gozi Trojan connection to C&C	21260	
		IRC_Botnet	650	
		Virut DNS request for C&C attempt	62512	
		Neris	718	85140
5	CTU-Malware-Capture-Botnet-50	Spyeye bot contact to C&C server attempt	423	
		Neris	245079	
		BLACKLIST DNS	36	
		IRC-bot sent spam BROWSER	81	245619
Total				836191

Note: The performance evaluation shows in Table 6, this rule has a high accuracy form Botnet attacks. All viruses that detected are the most serious attack on the network system and developing rapidly the attacks. However, these rule file is the best tool to help administrators.

Performance evaluation of fallibility. The fallibility is abnormal traffic which it is incorrectly Botnets detection as normal traffic. The values is lower that indicates better performance, shown in equations [12].

Table 7   The performance evaluation of Botnets_attack_3.rules file

No.	Rules Type	Botnets	Number of Alert	Total Detected Number
1	CTU-Malware-Capture-Botnet-42	Neris_Botnet sent to HTTP	308087	
		C&C_Botnets	159	
		IRC sent spam	9841	318087
2	CTU-Malware-Capture-Botnet-43	Ozdok botnet communication with C&C server attempt	347	
		Neris	169987	170334
3	CTU-Malware-Capture-Botnet-47	Spyeye bot contact to C&C server attempt	4189	
		C&C_Botnets	13569	
		BLACKLIST USER-AGENT	250	
		Win32.Domsingx.A contact to C&C	6181	24189
4	CTU-Malware-Capture-Botnet-49	Koobface worm submission of collected data to C&C ...	5874	
		Botnets attack Chinese_C&C_hosts on http	66169	
		Neris	23	72220
5	CTU-Malware-Capture-Botnet-50	Neris_Botnet sent to HTTP	227767	
		BLACKLIST DNS	35	
		IRC sent spam to HTTP	9875	
		TT-bot botnet contact to C&C server	204	237881
Total				822711

Note: The performance evaluation shows in Table 7, this is last of the rules for Botnets detection. The Botnets_attack_3.rules is including type of Botnet attack and it has the effective detection. This rule contains the information knowledge of the attack form attacker, this Botnets_attack_3.rules file can answer about every problem at the intrusion.

Table 8   Summary of Botnets detection of three files of rule

No.	Botnets Rules	Number of Botnets Detection
1	Botnets_attack_1.rules	809075
2	Botnets_attack_2.rules	836191
3	Botnets_attack_3.rules	822711

Note: The performances of evaluation shows in Table 8, each rules can be detected more efficient. The Botnets_attack_2.rules files can Botnet detection more than other rules. However, every rule files have important for Botnets detection.

Table 9   Efficiency of detection and fallibility performance of Botnets_attack_1.rules

No.	Datasets	Total Botnets of Datasets	Detected Number	Fallibility (%)	Efficiency of Detection (%)
1	CTU-Malware-Capture-Botnet-42	323154	296714	7.56	92.43
2	CTU-Malware-Capture-Botnet-43	176064	171373	2.66	97.33
3	CTU-Malware-Capture-Botnet-47	24764	24108	2.64	97.35
4	CTU-Malware-Capture-Botnet-49	85735	83235	2.91	97.08
5	CTU-Malware-Capture-Botnet-50	259949	233645	10.11	89.88

Note: In Table 9, efficiency of detection and fallibility performance of each datasets which we utilize Botnets_attack_1.rules file. Each datasets have value of Botnets detection different such as CTU-Malware-Capture-Botnet-47 can detection 97.35%, which this dataset is high efficiency for Botnets detection. Moreover, it has 2.64% of fallibility. In the other hand, the CTU-Malware-Capture-Botnet-50 can detection 89.88%, which this dataset is less efficiency of rules.

Table 10   Efficiency of detection and fallibility performance of Botnets_attack_2.rules

No.	Datasets	Total Botnets of Datasets	Detected Number	Fallibility (%)	Efficiency of Detection (%)
1	CTU-Malware-Capture-Botnet-42	323154	311384	3.64	96.35
2	CTU-Malware-Capture-Botnet-43	176064	169526	3.71	96.28
3	CTU-Malware-Capture-Botnet-47	24764	24522	0.97	99.02
4	CTU-Malware-Capture-Botnet-49	85735	85140	0.69	99.30
5	CTU-Malware-Capture-Botnet-50	259949	245619	5.51	94.48

Note: In Table 10 shows efficiency of detection and fallibility performance of Botnets_attack_2.rules file, which each of datasets have value high and low. The CTU-Malware-Capture-Botnet-49 can Botnets detection higher than other datasets, which it has efficiency of detection 99.30%. In addition, Malware-Capture-Botnet-49 is less of fallibility.

$$\text{Fallibility} = \frac{\text{FN}}{\text{FN} + \text{TP}} \times 100 \tag{1}$$

Performance evaluation of efficiency of detection. The efficiency of detection is indicative of the ability system, which the information has been required from detecting all of data. If a higher value, the system is effective in

Table 11   Efficiency of detection and fallibility performance of Botnets_attack_3.rules

No.	Datasets	Total Botnets of Datasets	Detected Number	Fallibility (%)	Efficiency of Detection (%)
1	CTU-Malware-Capture-Botnet-42	323154	318087	1.56	98.43
2	CTU-Malware-Capture-Botnet-43	176064	170334	3.25	96.74
3	CTU-Malware-Capture-Botnet-47	24764	24189	2.32	97.71
4	CTU-Malware-Capture-Botnet-49	85735	72220	15.76	84.23
5	CTU-Malware-Capture-Botnet-50	259949	237881	8.48	91.51

Note: In Table 11 shows efficiency of detection and fallibility performance of Botnets_attack_3.rules file, which the numbers of rules are lower "273" than other rules. The response of Botnets_attack_3.rules to the datasets reasonably well and the CTU-Malware-Capture-Botnet-42 is well for recompense. It can Botnets detection 98.43% of efficiency of detection. However, this CTU-Malware-Capture-Botnet-42 has value of fallibility in 1.56%.

Table 12   Summary for efficiency of detection and fallibility performance comparison

No.	Rule Files	Total of Fallibility (%)	Total Efficiency of Detection (%)
1	Botnets_attack_1.rules	5.17	94.81
2	Botnets_attack_2.rules	2.90	97.81
3	Botnets_attack_3.rules	6.27	93.72

Note: Efficiency of detection and fallibility of the performance comparison in Table 12, we observe the efficiency of detection on three rule files have difference efficiency to Botnet detection. The Botnets_attack_2.rules file is height efficiency of detection for Botnet detection, it's having 97.81%. Not only, it's having 2.90% of fallibility. In other hand, the Botnets_attack_3.rules of less efficiency of detection for detection, it is having 93.72% and it has value of fallibility 6.27%. However, the values of three rule files in efficiency of detection and fallibility are difference for Botnets detection, but it has helping administrator from attackers.

detection. In the other hand, if a low value that the system is effective a low detecting, shown in equations [12].

$$\text{Efficiency of detection} = \frac{TP + TN}{TP + TN + FP + FN} \times 100 \qquad (2)$$

where (TP) True positive is correctly identified of Botnet detection,
   (TN) True Negative is incorrectly identified of Botnet detection,
   (FP) False Positive is correctly rejected of Botnet detection,
   (FN) False Negative is incorrectly rejected of botnet detection.

6 Conclusion and Future Work

The Snort-IDS are effective intrusion detection and it is a network security tool which can monitor the abnormal behavior. This paper, we improve the Snort-IDS rules for Botnets detection and we analyze Botnets behaviors in three rule files. Moreover, we utilize the MCFP dataset, which includes five files such as CTU-Malware-Capture-Botnet-42, CTU-Malware-Capture-Botnet-43, CTU-Malware-Capture-Botnet-47, CTU-Malware-Capture-Botnet-49, and CTU-Malware-Capture-Botnet-50. The performance evaluation of Botnet detection, the performance evaluation of efficiency of detection, and the performance evaluation of fallibility for three rule files were evaluated by detecting each packet. The experimental results shown that, the case of Botnets_attack_1.rules file can maximally detect Botnets detection for 809075 alerts, the efficiency of detection and fallibility for Botnets detection are 94.81% and 5.17%, respectively. Moreover, in the case of Botnets_attack_2.rules file can detect Botnets up to 836191 alerts, having efficiency of detection and fallibility for Botnets detection are 97.81% and 2.90%, respectively. The last case Botnets_attack_3.rules file can detect Botnets 822711 alerts, it can 93.72% of efficiency of detection and the value of fallibility is 6.27%. However, the Botnets_attack_2.rules file is most proficient for Botnets detection, because it has a high efficiency of detection for Botnets detection and it has less of fallibility. Moreover, this paper can support the administrator to secure the network quickly. In addition, they must also regularly update the Snort-IDS rules, because the attackers could fine the way to attack the system any time. In the future work, we will find new techniques for intrusion detection system and using tool in data mining for more Botnets detection.

References

[1] Konhiatou, C. Y., Kittitornkun, S., Kikuchi, H., Sisaat, K., Terada, M., and Ishii, H. (2013). "Clustering Top-10 Malware/Bots based on Download Behavior," In *2013 International Conference On Information Technology and Electrical Engineering (ICITEE)*, Yogyakarta, 62–67.

[2] Ramsbrock, D., and Wang, X. (2013). *The Botnet Problem*, Chap. 12. Available at: http://www.sciencedirect.com/science/article/pii/B9780123 94397200012X

[3] García, S., Grill, M., Stiborek, J., and Zunino A. (2014). An Empirical Comparison of Botnet Detection Methods. *Comput Security*, 45, 100–123.

[4] Sathish, V., and Khader, P. S. A. (2014). Deployment of proposed bot-nets monitoring platform using online malware analysis for distributed environment. *Indian J. Sci. Technol.* 7, 1087–1093.

[5] Awadi, A. H. R. A. and Belaton, B. (2013). Multi-phase IRC botnet and botnet behavior detection model. *Int. J. Comput. Appl.* 66, 41–51.

[6] Li, W. M., Xie, S. L., Luo, J., and Zhu, X. D. (2013). A detection method for botnet based on behavior features *Adv. Mater. Res.* 765–767.

[7] Shah, S. N. and Singh, M. P. (2012). Signature-based network intrusion detection system using SNORT And WINPCAP. *Int. J. Eng. Res. Technol.* 1, 1–7.

[8] Geng, X., Liu, B., and Huang, X. (2009). "Investigation on secu-rity system for snort-based campus network," in *Proceedings of the 1st International Conference on Information Science and Engineering (ICISE)*, Nanjing University of Science and Technology, Nanjing, China, 1756–1758.

[9] Rani, S., and Singh, V. (2012). SNORT: an open source network security tool for intrusion detection in campus network environment. Int. *J. Comput. Technol. Electron. Eng.* 2, 137–142.

[10] Huang, C., Xiong, J., and Peng, Z. (2012). "Applied research on snort intrusion detection model in the campus network," in *IEEE Symposium on Robotics and Applications(ISRA)*.

[11] Roesch, M. (1999). "Snort – Lightweight Intrusion Detection for Networks," in *Systems Administration Conference*, Washington, USA, 229–238.

[12] Khamphakdee, N., Benjamas, N., and Saiyod, S. (2015). Improving Intrusion Detection System based on Snort Rules for Network Probe Attacks Detection With Association Rules Technique of Data Mining. *J. ICT Res. Appl.* 8, 234–250.

[13] http://mcfp.weebly.com/mcfp-dataset.html [Accessed May 2015].

[14] Khamphakdee, N., Benjamas, N., and Saiyod, S. (2014). "Improving Intrusion Detection System based on Snort Rules for Network Probe Attack Detection," in *International Conference on Information and Communication Technology (Icoict)*, Bandung, 69–74.

[15] Chanthakoummane, Y., Saiyod, S., and Khamphakdee N. (2015). "Eval-uation Snort-IDS Rules for Botnets Detection," In *National Conference on Infomation Technology.*

Biographies

S. Saiyod received the B.Sc. degree in Computer Science from Mahasarakham University in 2000 and M.Eng. degree in computer Engineering in 2005 and the D.Eng degree in 2011 form King Mongkut's Institute of Technology Ladkrabang, Thailand. His current interests are in the area of performance evaluation on communication networks, digital-signal-processing, and mobile communication.

Y. Chanthakoummane received his B.Eng. degree in Computer Engineering from National University of The Laos PDR in 2009. He is a master's student at the Department of Computer Science, Faculty of Science, Khon Kaen University, Thailand.

N. Benjamas received her B.Sc. degree in Computer Science at the Department of Computer Science, Faculty of Science, Khon Kaen University in 2000 and M.Sc. degree in Computer Science in 2005 and D.Eng. degree in Computer Engineering in 2012 from Kasetsart University, Thailand. Her current interests are big data analytics, data mining and knowledge discovery, parallel computing, distributed computing, high performance computing (HPC), and cloud computing.

N. Khamphakdee received his B.Sc. Computer Science from Udon Thani Rajabhat University in 2005 and M.Sc. degree in Computer Science in 2015 from Khon Kaen University. He is Ph.D student at Computer Science at the Department of Computer Science, Faculty of Science, Khon Kaen University. His current interests are in the area of data mining, big data, and network security.

J. Chaichawananit received his B.Sc. degree in Computer Science from Khon Kaen University, Thailand in 2014. He is a master's student at the Department of Computer Science, Faculty of Science, Khon Kaen University, Thailand.

Synflood Spoof Source DDOS Attack Defence Based on Packet ID Anomaly Detection – PIDAD

Tran Manh Thang and Van K. Nguyen

Dept of Software Engineering, School of Information Technology and Communication, Hanoi University of Science and Technology, No. 1 Dai Co Viet, Hai Ba Trung, Hanoi, Vietnam
E-mail: thang197@gmail.com; vannk@soict.hust.edu.vn

Received 2 April 2016; Accepted 13 May 2016;
Publication 3 June 2016

Abstract

A distributed denial-of-service (DDoS) attack characterized by flooding SYN packets is one of the network attacks to make the information system unavailable. This kind of attack becomes dangerous and more difficult to prevent and defense when attackers try to send flood SYN packets with spoof source, especially, there packets have information fields as the normal SYN packets. In this study, we propose a method called Packet Identification Anomaly Detection – PIDAD used to defense type of DDoS attack mentioned above. This method based on abnormal information of identification field in IP Header when observing the set of packets received in the victim system.

Keywords: DDoS attacks, DBSCAN, PIDAD.

1 Introduction

Distributed Denial-of-Service attack (DDoS) is a type of network attack that attackers make use of a large number of compromised computers (botnet) to attack, which makes the applications, service or an information system unavailable. This type of attacking has become more and more dangerous and

Convergence Security, Vol. 1, 41–56.
doi: 10.13052/jcs2445-9992.2016.003

harder to prevent when the number of computers connected to the Internet, the weakness of information security and malwares get increasingly going up. In fact, there have been many researches on how to prevent and defense against this kind of attack; however, it is shown that indeed there have been no effective measures of defense and resulting in serious impact on agencies, organizations, and business during the recent time.

DDoS attack comprises of various forms of attacks, wherein, TCP SYN Flooding [1] is one of the attacking types that is of most difficulty to be prevented and defensed. With this kind of attack, hackers control computers in botnet through controlling server (C&C Server) to send flooding spoof packets which have information fields like normal packets to target server. Therefore, servers attacked cannot differentiate which connection initiate packet is real.

One of the most specific, typical forms is TCP SYN Flood attack, amongst hackers aim at a handshaking process with 3 steps of initiating a connection TCP (TCP transport protocol of the Internet). During this process, hackers create flooding SYN packets sent to a victim server which incomplete the process of 3 step handshake. Thus, server has to spend resources for connections which are incomplete due to hackers' sending on purpose. This leads to server's running out of resources, incapable of meeting the requests of connecting for real connections.

To detect SYN spoof packets, we cannot check each SYN packet sent to server but we need to find out the relationship between them.

Basing on the principle of Idle scan method [26], which is used to check an open port on server basing on the characteristic of gradually increasing information field of Packet Identification – PID in the beginning part of IP packet (IP Header) when a packet is sent out of a computer, we see that it is possible to find out SYN spoof packets by searching for packets with information field of gradually increasing PID with different IP address source (hereinafter referred to as different IP address).

A matter here is that one computer can utilize many applications at the same time, conducting different procedures simultaneously, thus, packets with the same goals will be ones with interrupted gradually increasing PID field. To illustrate more clearly, supposing that a computer sending with 3 progresses A, B and C is at the same time communicating with 3 other computers. Then, if we overlook the very computer, PID stream is created continuously, but if observing PID stream sent from the progress A (or B, or C) to some receiving computer, it is obviously to increase interruptedly.

To detect spoof packets used in DDoS attack, we study the method of grossing packets with different source addresses with continuously increasing

PID value into the same Cluster using the algorithm DBSCAN [3]. As of each Cluster, we will define the expected value (Expected PID – EPID) of PID value each Cluster. Thence, any new packet send to the victim server will be considered as a spoof packet if PID value of new packet is equal EPID value in any Cluster.

In this paper, we will use some English specialized terms considered to be common in this area. In our opinion, using English terms with basic definition, common in international papers, will help specialized readers be easier to follow than translating into Vietnamese. We would like to state some English terms as below:

- Ddos: Distributed Denial-of-Service attack.
- SYN: packet in TCP protocol stack used initially to establish connection.
- PID: the value of Packet Identification field in IP Hearder.
- Cluster: cluster/the nearby object in the sample space.
- Training phase: computer training phase.
- Detection phase: detecting phase of spoofing packet.
- Core poin: point/object inside cluster.
- Border point: point/object on the border.

1.1 Field Overview

As the overview research of DDoS attack [16], the solutions of defense DDoS attack are applied in two main directions: deployment location and point in time that defense takes place.

Applying method basing on deployment location is divided into 2 classes of defense: the first class is defense solutions to DDoS attacks occurring at the Transport layer of OSI model downwards (network/transport-level DDoS flooding attacks). The second class is defense solutions at application layer of OSI model. With the defense method basing on point in time that defense takes place location is divided into 3 phases: pre-attack phase, attack phase, and post-attack phase.

For the solution of defense against DDoS attack at network/transport-level is divided into categories: source-based, destination-based, network-based, and hybrid and for the defense solutions against application level is divided into categories: destination-based, and hybrid based on their deployment location.

In this study, we focus on studying the method of defending DDoS attack type SYN flood spoofing source address. This is the solution applied for

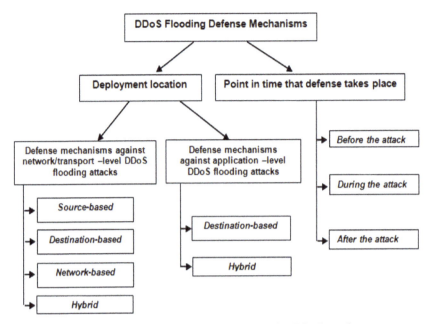

Figure 1   Categorize the method of defensing DDoS attacks.

defensing against DDoS attack at network/transport-level mentioned above. We will analyze in detail some related researches as following.

Method of IP Traceback mechanisms [17–18] is the method of saving information of the root from source to victim in the Packet Identification information field of IP packets corresponding to each source IP address. Basing on this information about this route to detect the spoof packets. Spoof packets are packets of route information mismatched its source IP stored beforehand. However, this method has a big limit which requires mediate router to support the mechanism of marking the route, thus it is difficult to carry out in reality.

Method of Management Information Base (MIB) [19–21] is the method of combining information existing in each packet and linear information to detect DDoS attack. When DDoS attack occur, this method compares the information in each packet of ICMP, TCP, UDP with the respective information which was analyzed and stored when DDoS attack had not occurred to find out the packets with abnormal information. Wang et al. [27, 28] proposed a method for detecting SYN flood attacks at leaf routers that connect end hosts to the Internet. Based on the observation that SYN and FIN packets form pairs in normal network traffic, they proposed using a nonparametric CUSUM method to accumulate the pairs.

Method of Packet marking and filtering mechanisms includes the following researches: History-based IP filtering [22], this method stores the information of source IP addresses which frequently connect to the system when DDoS attack has not occurred. When DDoS attack occurs, IP addresses existed in the list stored will be connected to the system; Method of Hop-count filtering [23], this method informs of source IP addresses as corresponding hop-count information in each packet which will be stored when DDoS attack has not occurred. When DDoS attack occurred, the packets with source addresses stored beforehand while information about hop-count different from information stored beforehand relating with that source IP addresses will be considered as spoof; Method of Path Identifier (Pi) [24], this method stores the values which are regarded as identifying the route of each packet when traveling from source to victim corresponding to each source IP address. The packets with the same route to the victim will share the same route information. This information will be used to filter all spoof packets of the same route with a spoof packet detected beforehand.

Method of Packet dropping based on the level of congestion wherein Packetscore [25] is a typical research for this method. Packetscore sets up priority level for each packet basing on the algorithm of Detecting-Differentiating-Discarding routers (3D-R) using Bayesian-theoretic metric. A barrier will be set up to filter the packets with low priority level when DDoS attack occurs and bring about stuck at Routers.

2 Theoretical Basis

2.1 Packet Identification (PID)

This is Identification field (16 bits) in IP Header [2]. This field is used to determine each packet sent by a computer. By default, the value of PID field will be increased by 1 unit when the computer sends out a packet. This characteristic has been used in the method of Idle Scan to check one port opened in server [26].

2.2 TCP Handshake and TCP SYN Flooding Attack

Transmission Control Protocol (TCP) [4] is protocol used popularly in the Internet to create a trustful connection oriented between any two computers in the Internet. Wherein, TCP handshake is a mechanism used to initiate a new TCP connection. TCP Handshakes can be described briefly as following: In the process of initiate new TCP connection, first, client sends a packet with

SYN flag to server to require initiating new connection. After receiving the requirement, server will response a packet with SYN-ACK flags to inform the readiness for connection and require initiating new connection to client. The last step, client also responses a packet with ACK flag to server informing its readiness. Because this protocol has been developed from early time (when the modern Internet has not formed, the potential risks of Internet attacks have been unknown), it can be said that this connection principle is a dangerous weakness of information security that attackers can make use of it to perform different kinds of attacks.

The weakness here is: when receiving requirement of initiating new connection from client, immediately server reserve the necessary resources for new connection. Taking advantage of this loose, attacker will try to send overwhelm spoof packets and never complete TCP handshakes progress so that the server gradually depletes its resources.

Spoof packets are created by randomly spoofing source IP address and other information in each spoof packet is completely the same as normal packet. This leads to: server cannot differentiate which connection packet which packet is the real or spoof one, so server will spend resources for such unreal connections. As the number of connection spoof packets is flooded sending, server has no longer had resources to serve real connections.

2.3 DBSCAN Algorithm

DBSCAN (Density-based spatial clustering of applications with noise) [29] is a data clustering algorithm proposed by Martin Ester, Hans-Peter Kriegel, Jörg Sander and Xiaowei Xu in 1996. It is a density-based clustering algorithm:

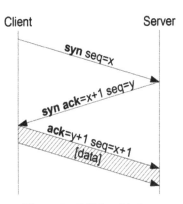

Figure 2   TCP handshakes.

given a set of points in some space, it groups together points that are closely packed together (points with many nearby neighbors), marking as outliers points that lie alone in low-density regions (whose nearest neighbors are too far away). DBSCAN is one of the most common clustering algorithms and also most cited in scientific literature.

The main idea of the algorithm is basing on bringing up *core point* term (core point, lying inside one cluster) – which has the number of neighbors reached standard level in a given neighboring radius – and the forming of clusters as spreading type based on the definition of connection or reachable destiny. In each cluster formed as DBSCAN, point objects are categorized into 2 types: *core point* (as the real object inside cluster) and *border point* (object lying on the border of cluster). We will present briefly summary of basic definitions and operative mechanism of DBSCAN algorithm as following.

2.3.1 Definitions

Neighboring radius area ε of object p, symbol $N_\varepsilon(p)$, is a collection of object q under condition that the distance between p and q, symbolized as *dist*(p,q), is smaller than given radius $\varepsilon > 0$:

$$N_\varepsilon(p) = \{q \in \mathrm{D} | dist(p,q) \le \varepsilon\} \tag{1}$$

An object can be considered to have thick or thin neighboring density through the size of this collection $N_\varepsilon(p)$ We also define the term *reachable* density-based (directly density-reachable) from p to q: p regarded as (ε,m)-*reachable directly* with q if $p \in N_\varepsilon(q)$ and $|N_\varepsilon(q)| \ge m(m$ usually symbolized as *minPts* in the completed documents on DBSCAN). Generally speaking, this concept shows that p can be reachable to q thanks to locating nearby and there are

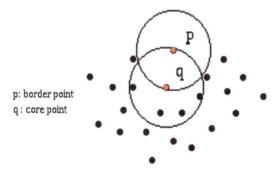

Figure 3   DBSCAN algorithm.

many mediate points within the neighborhood of *q*. The relationship *directly density-reachable* can be on one side or two sides (or symmetry). If it is two sides (p reachable to q and vice verse), both two points are remarked as core point, but if only *p* is reachable to *q*, q then is remarked as *core* and p is *border point*.

In overall, object *p* is considered (*ε,m*)-*reachable* with *q* if there is string $p_1, p_2, \ldots, p_n$ ($p_1 = q$, $p_n = p$) existing while p_{i+1} directly density-reachable to p_i: $p_{i+1} \in N_\varepsilon(p_i)$ và $| N_\varepsilon(p_i)| \geq m$.

Object *p* connected as the density, symbolized as (*ε,m*)-*connected*, wit object *q* if there is object *o* existing while both *p* and *q* are density-reachable, i.e.,(*ε,m*)-*reachable,* from *o*.

A cluster satisfies the standard of density (*ε,m*)-*dense* if all of its objects are (*ε,m*)-*connected* with one another and any object (*ε,m*)-*reachable* from some object of cluster is also considered to belong to cluster.

2.3.2 Algorithm

DBSCAN uses two parameters set up before which are *ε* and *m* = *minPts,* from which searching detecting (actually building and forming) *cluster*s under the condition of density (*ε,m*)-*dense*. Algorithm of DBSCAN begins from any object and then searches every object which is (*ε,m*)-*reachable* from object *p*. If *p* is *core point*, this task will generate a cluster satisfying (*ε,m*)-*dense*. If *p* is border object, it will be impossible to search an object density-reachable from *p*, then DBSCAN scan the next object in the database. Due to using the sharing parameter *ε* and *m* for all of the Cluster, DBSCAN can combine 2 clusters into 1 if these two clusters are nearby. The distance between 2 collections of S1 and S2, symbolized as *dist* (S1, S2) is the value of the smallest distance of any 2 objects *p*, *q* insides: *dist* (S1, S2) = min{*dist*(p,q) | p ∈ S1, q ∈ S2}. Two collections of objects in the same cluster can be separate if their distance between them increases.

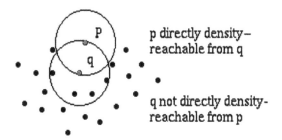

Figure 4   Directly density-reachable Relationship.

3 Pidad Method

PIDAD method we suggested is based on the main idea detecting strings of packets with their PID value continuously increasing but source addresses are random, without coincidence. As we mentioned above (Introduction part) the packets created and sent from the same computer will have PID value increase continuously as short section; these continuously increasing sections make up the sections without coincidence and somehow interrupted. The reasons as stated are that in the same computer there can be many progresses having transactions connected to many other computers in the Internet. However, in TCP SYN flooding attack, attackers are supposed to create spoof SYN packets with spoof source IP address selected randomly. Therefore, we suggest a method to detect and filter packets, which have PID value continuously increasing short sections with different source IP addresses (due to random generation).

To realize this idea, we recommend using DBSCAN algorithm: each PID string with increasing short sections of different source IP addresses will be collected as a cluster as DBSCAN. With each cluster, we will determine E_{PID} value of each cluster, i.e., PID value is predicted to be sent continually. The process of forming clusters and find out E_{PID} in our method is called *Training phase*.

To detect the next spoof packets sent to the system, we will check PID value of each packet sent to see if there is coincidence with E_{PID} of any Cluster (list of E_{PID} collected through *training phase*); in case of finding PID value coinciding with any E_{PID} value, this packet will be the next spoof packet sent to the system. In case of unable to find PID value which coincide with any E_{PID} value, the packet sent can be real packet or the first spoof packet of a new Cluster (unformed in *training phase*).

We also suggest a technique using another restricted condition of time to build a new Cluster. Beside the phenomenon of PID continuously increasing short sections as mentioned above, normally the spoof packets sent from a minion computer in DDoS attack will be rather nearby and regular in terms of time (as it is necessary to create a large number of spoof packet without attention to responses from victim server). Thus, we suggest another condition of forming Cluster: in the time period of $T_{CC} = 2*T_{PC}$ there must be at least 2 coming packets sent and having full conditions to create new Cluster with the packet needed to be checked, then the new Cluster will be set up with the respective E_{PID} value. Wherein T_{PC} is the maximum time period of each Cluster needed to add a new member. If during the time period T_{CC} it is under conditions to set up new Cluster to the packet being checked, that packet will be considered to be real packet. The process of detecting next spoof packet sent to the system in our method called *Detection phase*.

As above, we have described the basic mechanism of the suggested algorithm; we would like to demonstrate each phase of the algorithm in detail as following.

3.1 Pidad Training Phase

To have enough input sample for DBSCAN algorithm in training phase, first PIDAD method needs to collect N_{pc} initial packets with different source IP

```
For each Pi in MC
  if (Sp(Pi) is notcheck)
      set CorePoint = Pi;
      // Scan to get new member
      for each Pk (j < k < Npc)
          //check BorderPoint P(k)
          if ((Trc(Pj) - Trc(Pi) < Tpc) and (PID(Pj) == PID(Pi) + 1))
              set CorePoint = Pj
              for each Pk (j < k < Npc)
                  //check BorderPoint P(k)
                  if ((Ttc(Pk) - Ttc(Pj) < Tpc) and (PID(Pk) == PID(Pj) + 1))
                      Create new Cluster CT
                      add CT infor to Mc
                      set CorePoint = Pk;
                      set EPID = PID(Pk) + 1;
                      add Pi,Pj,Pk infor to Mpc
                      set Sp(Pi),Sp(Pj),Sp(Pk) = checked
                  end
              end
          end
      end
end
```

Figure 5   PIDAD algorithm in training phase.

addresses sent to the system. After having N_{pc} packets, PIDAD will apply DBSCAN algorithm to find out Clusters which have packets with continuously increasing PID. To describe the characteristic of continuously increasing by 1 unit of each Cluster, DBSCAN algorithm chooses parameter ε (Eps) = 1 (the distance between 2 points in one Cluster). To describe the characteristic of the packets sent from the same computer the value MinPts = 3 (this value has the meanings of at least 3 packets with continuously increasing PID creating one Cluster). In case the value MinPts = 2 the estimate PID value of this Cluster miscollected into other Cluster will be of high potential, which reduces the accuracy of the method.

Due to characteristic of directly density-reachable relationship $((\varepsilon,m)$- reachable) of DBSCAN algorithm, whenever Cluster has new member (a new packet satisfied the conditions and belonging to that Cluster) the position of core point will be updated into the position of that new member (border point).

PIDAD uses a multi-dimension matrix M_C to be able to store information of Clusters including: value of Cluster ID, E_{PID} and the time Cluster add the last member (T_{LC}). Using M_{PC} to store the information of each packet sent to the system including PID, Cluster ID and S_P indicates the state of whether the packets checked become member of Cluster or not. Supposed P_i the packet i ($0 < i < N_{pc}$) which is checked of all conditions to bring into a Cluster, T_{TC} is the time to check packet P_i. Algorithm and algorithm flowchart in training phase are following.

3.2 Pidad Detection Phase

After setting up Clusters and E_{PID} correspond from N_{PC} packets in training phase, PIDAD will transfer to detection phase, to detect the next spoof packets sent to the system. Each next packet sent to the system can be spoof packet or real packet. Spoof packets are packets with PID coincident with any E_{PID} value in M_C. The real packets are the packets satisfying both two conditions at the same time: no PID value coinciding with any E_{PID} value and under conditions to create one Cluster in a period of time T_{CC} (after the time period T_{CC} the packets being checked cannot collect the next two packets to make a new Cluster). T_{CC} value set up is smaller than TCP timeout to allow PIDAD to determine the real packets before occurring TCP retransmission at Client side.

To check if a packet has enough conditions to make a new Cluster or not, PIDAD method uses a temporary Cluster (C_T). Each packet sent to the system under the state of being checked will be assigned to be core point of C_T, after a time period T_{CC}, when C_T collects 2 border points C_T will be updated into new Cluster, or else C_T will be deleted. MT_C is used to store the information of C_T, the information needed to be stored of C_T as well as information of Cluster was set up in training phase. Supposed N_{PT} as the number of packets in each C_T.

In detection phase, to meet the ability of detecting spoof packets when having changes of attacking flow and minimizing the number of Clusters needed to be handled, PIDAD method should have updating method, deleting and adding new Clusters. Wherein, a Cluster is updated E_{PID} information when that Cluster add another new member. A Cluster will be deleted if after the time T_{CC}, that Cluster does not have new member added. A Cluster will be new created when after the time T_{CC} the packets being checked collect the next 2 packets with full conditions to set up a new Cluster.

Supposed P_i the packet number i being checked, $C(p)$ is Cluster with p as core point. Algorithm in detection phase is described as the following figure:

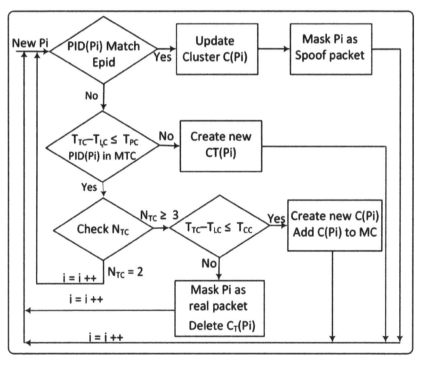

Figure 6   PIDAD algorithm in detection phase.

4 Experiment Evaluation

The experimental results in this study, we take the sample of packets with real time in some real systems when meeting DDoS attack with the above type of attack. To evaluate the effectiveness of the method more accurately, we have collected connection initial packets when the system work at normal mode and mixed randomly with the sample packets into experimental data. Experimental data are stored as file PCAP. Wherein the packets at normal functional mode and the sample packets during the attack will be marked to have basis of calculating the efficiency of the algorithm.

To collect SYN packets in file PCAP, we use the software Wireshack [14] to filter SYN packets in PCAP. To have information on the arrival time of the packets, and PID information from the file SYN packets collected, we use the software Tshark [15] to filter two information fields time_stemp and identification of each packet.

After having information of arrival time and PID of files of packets and programming the algorithm on the software Matlab, the experimental results show that our method can detect most of spoof packets sent to the system when meeting DDoS attack. The specific results are following.

5 Future Works

In this research, we have studied and suggested a new method to detect spoof packets used in DDoS attack. The results show that our method has high rate of detecting spoof packets. However, in this research we still have to conduct manually some steps of solution in some parts such as: collecting initial packets with different IP in N_{PC}; the storing information of the packets was confirmed and connected initial successfully. To be able to complete and apply the method in reality, we anticipate to doing research on using Bloom filter algorithm suggested by Bloom [3] in the future study, this algorithm was used in some methods of preventing, defensing from DDoS attack [5–7]. Specifically, we will continue to study using Bloom filter algorithm to determine the first SYN packets sent to the system, to store information of connection initiated and store EPID value of each Cluster.

Table 1   Experimental results of PIDAD

N_{pc}/T_{pc}	0.1 ms	0.2 ms	0.3 ms	0.4 ms
10000	86.2%	92.1%	84.1%	76.1%
20000	87.1%	91.1%	8.1%	77.1%
30000	88.1%	89.1%	94.1%	79.1%
40000	85.1%	95.1%	88.1%	80.1%

6 Conclusion

PIDAD method allows detecting and filtering connection initiate spoof packets used in DDoS attack typed flooding connection spoof packets. This method has advantages of allowing detecting spoof packets sent to the system without requiring Client side to resend connection initiate packets the second time. Basing on experimental results, it shows that this method can detect most of the spoof packets in DDoS attack. However, in this research, we just focus on the method of detecting spoof packets without having recommendation of overall solution to completely apply in reality. In future studies, we need to do research using Bloom filter algorithm to complete and improve the troubleshooting efficiency of the proposed method.

References

[1] CERT. *TCP SYN Flooding and IP Spoofing Attacks*. Advisory CA-96.21, September 1996.

[2] http://en.wikipedia.org/wiki/IP_header

[3] Ester, M., Kriegel, H. P., Sander, J., and Xu, X. (1996). "A density-based algorithm for discovering clusters in large spatial databases with noise," in *Proceedings of the 2nd International Conference on Knowledge Discovery and Data Mining*, Portland, Oregon, 226–231.

[4] Postel, J. (1981). *Transmission Control Protocol: DARPA internet program protocol specification, RFC 793*.

[5] Abdelsayed, S., Glimsholt, D., Leckie, C., Ryan, S., and Shami, S. (2003). "An efficient filter for denial-of-service bandwidth attacks," in *IEEE Global Telecommunications Conference (GLOBECOM'03)*, Vol. 3, 1353–1357.

[6] Snoeren, A. C. (2001). "Hash-based IP traceback," in *Proceedings of the ACM SIGCOMM Conference*, (Boston, MA: ACM Press), 3–14.

[7] Yaar, A., Perrig, A., and Song, D. (2003). "Pi: a path identification mechanism of defend against DDoS attacks," in *IEEE Symposium on Security and Privacy*, 93.

[8] Yaar, A., Perrig, A., and Song, D. (2003). *StackPi: New Packet Marking and Filtering Mechanisms for DDoS and IP Spoofing Defense, CMU-CS-02-208*.

[9] Chen, W., and Yeung, D. Y. (2006). "Defending against TCP SYN flooding attacks under different types of IP spoofing," in *Fifth International Conference on Networking (ICN)*.

[10] Changhua, S., Jindou, F., Lei, S., and Bin, L. (2007). "A novel router-based scheme to mitigate SYN flooding DDoS attacks," in *IEEE INFOCOM (Poster)*, Anchorage, Alaska, USA, 2007.

[11] Chan, E., Chan, H., Chan, K., Chan, V., Chanson, S., et al. (2004). "IDR: an intrusion detection router for defending against distributed denial-of-service(DDoS) attacks," in *Proceedings of the 7th International Symposium on Parallel Architectures, Algorithms and Networks 2004(ISPAN'04)*, 581–586.

[12] Wang, H., Jin, C., and Shin, K. G. (2007). Defense against spoofed IP traffic using hop-count filtering. *IEEE/ACM Trans. Netw.*, 15, 40–53.

[13] Peng, T., Leckie, C., and Ramamohanarao, K. (2003). "Protection from distributed denial of service attacks using history-based IP filtering," in *ICC'03*. Vol. 1, 482–486.

[14] https://www.wireshark.org/

[15] https://www.wireshark.org/docs/man-pages/tshark.html

[16] Taghavi Zargar, S., and Tipper, D. (2013). A survey of defense mechanisms against distributed denial of service (DDoS) flooding attacks. *IEEE Commun. Survey Tutorials*, 15, 2046–2069.

[17] John, A., and Sivakumar, T. (2009). DDoS: survey of traceback methods. *Int. J. Recent Trends Eng.* 1.

[18] Joao, B., Cabrera, D., et al. (2001). "Proactive detection of distributed denial of service attacks using MIB traffic variables a feasibility study," in *Integrated Network Management Proceedings*, Seattle, WA, 609–622.

[19] Jalili, R., and ImaniMehr, F. (2005). *Detection of Distributed Denial of Service Attacks Using Statistical Pre-Prossesor and Unsupervised Neural Network, ISPEC*, 192–203. Berlin: Springer-Verlag.

[20] Li, M., Liu, J., and Long, D. (2004). *Probability Principle of Reliable Approach to detect signs of DDOS Flood Attacks, PDCAT*, 596–599. Berlin: Springer-Verlag.

[21] Wang, H., Jin, C., and Shin, K. G. (2007). Defense Against Spoofed IP Traffic Using Hop-Count Filtering, *IEEE/ACM Trans. Netw.* 15, 40–53.

[22] Yaar, A., Perrig, A., and Song, D. (2003). "Pi: A Path identification mechanism to defend against DDoS attacks," in *IEEE Symposium on Security and Privacy*, pp. 93.

[23] Kim, Y., Lau, W. C., Chuah, M. C., and Chao H. J. (2006). PacketScore: a statistics-based packet filtering scheme against distributed denial-of-service attacks. *IEEE Trans. Depend. Secure Comput.* 3, 141–155.

[24] https://en.wikipedia.org/wiki/Idle_scan

[25] Wang, H., Zhang, D., and Shin, K. G. (2002). "Detecting SYN flooding attacks," in *Proceedings of Annual Joint Conference of the IEEE Computer and Communications Societies(INFOCOM)*, Vol. 3, 1530–1539.

[26] Wang, H., Zhang, D., and Shin, K. G. (2004). Change point monitoring for the detection of dos attack. *IEEE Trans. Depend. Secure Comput.* 1, 193–208.

[27] Ester, M., Kriegel, H. P., Sander J., and Xu, X. (1996) "A density-based algorithm for discovering clusters in large spatial databases with noise," in *Proceedings of the 2nd International Conference on Knowledge Discovery and Data Mining*.